Network as a Service Transforming Business Connectivity

By- Raghav Kathuria

Index

Chapter 2: Technical Foundation of NaaS- Making Sense of the Technology

Chapter 3: NaaS Service Models

Chapter 4: Implementing NaaS in Enterprises

Chapter 5: NaaS Security and Compliance

Chapter 6: Building a Business Case for NaaS

Chapter 7: Economic Implications of NaaS

Chapter 8: Future Trends in NaaS

Chapter 9: Industry Specific NaaS Application

Chapter 10: NaaS Challenges and Limitations

Chapter 11: Conclusion and Future Outlook

Appendices

Network as a Service Transforming Business Connectivity

Preface

In an era where digital transformation is not just an advantage but a necessity, the way enterprises connect and communicate has undergone a fundamental shift. Network as a Service (NaaS) has emerged as a revolutionary paradigm that promises to transform how businesses approach their networking infrastructure, security, and connectivity needs.

As we stand at the crossroads of traditional networking and cloud-native solutions, enterprises face critical decisions about their digital infrastructure. The acceleration of remote work, the proliferation of IoT devices, and the increasing adoption of multi-cloud strategies have created unprecedented demands on corporate networks. Traditional networking approaches, with their rigid architectures and capital-intensive models, are struggling to keep pace with these evolving requirements.

This book arrives at a crucial moment in enterprise networking history. It serves as both a comprehensive guide and a strategic framework for business leaders, IT professionals, and network architects who are navigating the transition to Network as a Service. Whether you're a CTO evaluating the business case for NaaS adoption, a network engineer designing hybrid cloud solutions, or an IT manager planning for scalable and secure connectivity, this book provides the insights and practical knowledge needed to make informed decisions.

Throughout these pages, we explore the fundamental principles of NaaS, from its technical foundations to its business implications. We examine real-world case studies of successful NaaS implementations, analyse the challenges and opportunities in various industry contexts, and provide actionable frameworks for assessment and adoption. Special attention is given to critical aspects such as security, compliance, and integration with existing infrastructure – concerns that are paramount in enterprise environments.

The book is structured to take readers on a journey from understanding the basics of NaaS to implementing sophisticated enterprise-wide solutions. Each chapter builds upon the previous one, offering a blend of theoretical knowledge and practical applications. We've included detailed technical specifications where necessary, while keeping the overall narrative accessible to business decision-makers who may not have deep technical backgrounds.

One of the unique aspects of this book is its forward-looking perspective. While we thoroughly cover current NaaS offerings and implementations, we also explore emerging trends and technologies that will shape the future of enterprise networking. From AI-driven network optimization to edge computing integration, we examine how NaaS will evolve to meet tomorrow's challenges.

As you embark on this exploration of Network as a Service, I encourage you to approach the material with both a critical mind and an open perspective. The transformation of enterprise networking is not just about technology – it's about enabling businesses to become more agile, resilient, and innovative in an increasingly connected world.

Whether you're considering your first steps into NaaS or looking to optimize your existing network infrastructure, this book aims to be your trusted companion in understanding and leveraging the power of Network as a Service for your enterprise.

Welcome to the future of enterprise networking.

Raghav Kathuria

October 2024

Chapter 1:

Introduction to Network as a Service (NaaS)

1.1 The Evolution of Enterprise Networking

Enterprise networking has come a long way from its humble beginnings of simple local area networks (LANs) connecting a few computers. In the early days, organizations built and maintained their own physical network infrastructure, investing heavily in hardware, dealing with complex configurations, and managing ongoing maintenance. This traditional approach, while revolutionary for its time, is increasingly ill-suited for today's dynamic business environment.

The journey from traditional networking to Network as a Service (NaaS) can be traced through several key phases:

1. **Traditional Networking (1980s-1990s)**

   - Physical infrastructure owned and maintained by organizations

   - Capital-intensive investments in hardware

   - Limited scalability and flexibility

- High maintenance overhead

2. **Managed Services Era (2000s)**

   - Outsourcing of network management

   - Service Level Agreements (SLAs)

   - Professional network monitoring

   - Still hardware-dependent

3. **Cloud Revolution (2010s)**

   - Virtualization of network resources

   - Software-defined networking

   - Network function virtualization

   - Hybrid cloud environments

4. **NaaS Emergence (2020s)**

   - Consumption-based networking

   - Cloud-native architecture

   - API-driven management

   - Zero-trust security integration

1.1.1 Industry Example: Retail Transformation

Target Corporation's network evolution serves as a prime example of this transformation. In 2019, they began migrating from traditional networking to a NaaS model for their 1,800+ stores:

- **Before NaaS:**

  - Fixed bandwidth for all stores

  - Long deployment times for new locations

  - Complex management requirements

  - Limited visibility into performance

- **After NaaS:**

  - Dynamic bandwidth allocation

  - Rapid store connectivity

  - Simplified management

  - Enhanced performance monitoring

  - 40% reduction in networking costs

1.2 Defining Network as a Service

Network as a Service represents a fundamental shift in how organizations consume and manage network resources. At its core, NaaS is a cloud service model where network infrastructure, operations, and security are delivered as a service on a subscription basis. This model aligns with other "as-a-service" offerings that have transformed IT infrastructure, such as Software as a Service (SaaS) and Infrastructure as a Service (IaaS).

1.2.1 Key Characteristics of NaaS:

1. **Subscription-Based Model**

  - Pay-as-you-go pricing

  - Operational expenditure vs. capital expenditure

  - Flexible scaling options

  - Usage-based billing

2. **Cloud-Native Architecture**

  - Distributed infrastructure

  - Containerization

  - Microservices

  - API-first design

3. **Automated Operations**

  - Self-service provisioning

  - Automated scaling

- AI-driven optimization

- Predictive maintenance

4. **Integrated Security**

- Zero-trust architecture

- Built-in threat protection

- Compliance monitoring

- Identity management

1.3 The Business Case for NaaS

Organizations are increasingly turning to NaaS for several compelling reasons:

Financial Benefits

- Reduced capital expenditure

- Predictable operational costs

- Lower total cost of ownership

- Improved resource utilization

Operational Advantages

- Faster deployment times

- Reduced complexity

- Enhanced visibility

- Improved reliability

Strategic Value

- Business agility

- Innovation enablement

- Competitive advantage

- Risk mitigation

1.3.1 Industry Example: Healthcare Provider

Cleveland Clinic implemented NaaS across 18 hospitals and 220 outpatient locations, achieving:

- 99.999% network availability

- 60% reduction in network incidents

- 45% improvement in application performance

- 30% reduction in operational costs

- Enhanced patient care delivery

- Improved telemedicine capabilities

1.4 Core Components of NaaS

A comprehensive NaaS solution typically includes several key components:

1. **Network Infrastructure**

   - Virtual and physical resources

   - Network functions

   - Software-defined networking

   - Edge computing capabilities

2. **Management Platform**

   - Centralized control

   - Analytics and monitoring

   - Policy management

   - Configuration automation

3. **Security Services**

   - Threat detection and response

   - Access control

   - Encryption

   - Compliance management

4. **Service Level Agreements**

- Performance guarantees

- Availability commitments

- Support requirements

- Compliance standards

1.5 Common Use Cases

NaaS adoption is driven by various business needs:

Remote Workforce Enablement

- Secure remote access

- Consistent user experience

- Global connectivity

- Device management

Cloud Connectivity

- Multi-cloud networking

- Cloud on-ramp services

- Hybrid cloud integration

- Cloud security

Branch Networking

- SD-WAN services

- Branch security

- Local breakout

- Application optimization

Digital Transformation

- API integration

- DevOps enablement

- IoT connectivity

- Analytics capabilities

1.6 Challenges and Considerations

While NaaS offers numerous benefits, organizations must carefully consider several factors:

Technical Challenges

- Legacy system integration

- Migration complexity

- Performance requirements

- Technical expertise

Business Considerations

- Cost analysis

- Vendor selection

- Contract terms

- Change management

Security and Compliance

- Data protection

- Regulatory requirements

- Risk assessment

- Audit capabilities

1.7 Looking Ahead

As we delve deeper into subsequent chapters, we will explore each aspect of NaaS in greater detail, providing practical guidance for:

- Assessing NaaS readiness

- Selecting appropriate solutions

- Planning implementations

- Managing transitions

- Optimizing operations

- Ensuring security and compliance

Understanding these foundational concepts is crucial for organizations considering or implementing NaaS solutions. The following chapters will build upon this foundation, offering detailed insights and practical recommendations for successful NaaS adoption.

1.8 Chapter Summary

This chapter has introduced the fundamental concepts of Network as a Service, including:

- The evolution from traditional networking to NaaS

- Key characteristics and components of NaaS solutions

- Business benefits and use cases

- Important considerations for adoption

- Framework for future exploration

In Chapter 2, we will focus on technical foundation of NaaS and examine the architecture of NaaS in detail, including service models and integration patterns.

Chapter 2

Technical Foundations of NaaS- Making Sense of the Technology

2.1 Introduction: Why Understanding the Technology Matters

Imagine you've just moved into a new house. You don't need to know exactly how the electrical wiring works, but understanding the basics helps you use the system effectively, know when something's wrong, and communicate with electricians when needed. The same principle applies to Network as a Service (NaaS). While you don't need to be a network engineer, understanding the fundamental building blocks will help you make better decisions for your organization.

2.2 The Building Blocks: Think of It Like a Smart City

2.2.1 The Brain of the Network (Control Plane)

Think of the control plane as the city's traffic control center. Just as traffic controllers monitor and direct vehicles throughout the city, the control plane manages data flowing through your network.

Real-World Example:

Consider Sarah, an IT manager at a retail chain with 500 stores. Before NaaS, when she needed to prioritize traffic for the holiday shopping season, she had to manually configure each store's network equipment - a process that took weeks. With NaaS's control plane, Sarah can now make this change for all stores with a few clicks from a central dashboard, just like a traffic controller adjusting all traffic lights in a city from one control room.

2.2.2 The Roads and Vehicles (Data Plane)

If the control plane is the traffic control center, the data plane is like the actual roads, vehicles, and traffic lights that move people and goods around the city.

Real-World Example:

Take Netflix streaming service. When you press play, the data plane ensures your movie data travels from Netflix's servers to your device efficiently, adjusting for congestion (like a GPS suggesting alternate routes during rush hour) and maintaining quality (like ensuring emergency vehicles get priority on roads).

2.2.3 The Management System (Management Plane)

This is like the city planning office that monitors everything, collects data, and plans improvements.

Real-World Example:

James, a CIO at a manufacturing company, used to struggle with network outages that disrupted production. The management plane now alerts him before problems occur, just like how modern cities use sensors to predict and prevent traffic jams or power outages.

2.3 Making It Work: The Core Technologies

2.3.1 Software-Defined Networking (SDN): The Smart Traffic System

Imagine replacing all traffic cops with a smart traffic system that automatically adjusts to conditions. That's what SDN does for networks.

Before SDN:

A bank with 100 branches needed to send IT staff to each location to make network changes. It was like having a traffic cop at every intersection who couldn't communicate with others.

After SDN:

The same bank now controls all branch networks from headquarters. When they launch a new service, they can update all branch networks instantly - like a smart traffic system adjusting all traffic lights citywide for a major event.

2.3.2 Network Function Virtualization (NFV): The Swiss Army Knife

Instead of having different physical devices for each network function, NFV puts everything into software, like having a smartphone app replace multiple physical tools.

Real-World Example:

A hospital used to have separate physical devices for their firewall, load balancer, and router at each location. With NFV, they now run all these functions as software on standard servers, reducing costs by 60% and making updates as simple as installing a smartphone app.

2.3.3 Security: Building a Modern Castle

The Zero Trust Approach: Trust No One, Verify Everyone

Traditional security was like a castle with strong walls but free movement inside. Zero Trust is like having security checks everywhere, even inside the castle.

Before Zero Trust:

- Once someone got through the company VPN, they had access to most resources

- A compromised laptop could access multiple systems

- Security breaches were hard to contain

After Zero Trust:

- Every access request is verified, regardless of location

- A compromised device can only access limited resources

- Security breaches are contained quickly

Real Success Stories

Global Retail Chain Transformation

A major retail chain with 2,000 stores implemented NaaS:

Before:

- Store network outages took 4-6 hours to fix

- New store setup took 3 weeks

- Black Friday network crashes were common

- Security updates took months to implement

After:

- Outages automatically fixed in minutes

- New stores connected in 2 days

- No more holiday season crashes

- Security updates applied instantly across all stores

- Cost savings: $4 million annually

Healthcare Provider's Journey

A healthcare network with 15 hospitals and 100 clinics adopted NaaS:

Before:

- Telemedicine services were unreliable

- Patient record access was slow

- Network changes required weekend maintenance

- Security was a constant concern

After:

- Seamless telemedicine experience

- Instant access to patient records

- Changes made during business hours with no disruption

- Enhanced security and HIPAA compliance

- Patient satisfaction improved by 45%

2.4 Practical Tips for Your NaaS Journey

2.4.1 Starting Small

1. Begin with a pilot project

   Example: One regional office or a single department

2. Focus on quick wins

   Example: Automating routine network changes that currently take hours

3. Measure and share success

   Example: Track time saved, cost reduced, and user satisfaction

2.4.2 Building Momentum

1. Document improvements

   Example: "Network changes that took 2 weeks now take 2 hours"

2. Train your team

   Example: Regular lunch-and-learn sessions about new capabilities

3. Communicate wins

   Example: Monthly updates to stakeholders with concrete benefits

2.4.3 Looking to the Future

Emerging Technologies

Think of these as upcoming city improvements:

1. Artificial Intelligence

Like having a predictive system that knows you'll need more network capacity next Tuesday because of scheduled video conferences

2. Edge Computing

Like having mini distribution centers throughout a city instead of one central warehouse

Conclusion: Making It Work for You

Remember, just as you don't need to be a civil engineer to benefit from well-designed roads, you don't need to be a network expert to benefit from NaaS. The key is understanding enough to make informed decisions and maximize the value for your organization.

Next Steps:

1. Assess your current network needs

2. Identify pain points that NaaS could solve

3. Start small with a pilot project

4. Measure and communicate success

5. Expand based on proven results

Chapter 3:

NaaS Service Models

Network-as-a-Service (NaaS) is transforming how enterprises design, deploy, and manage their network infrastructure. By shifting from traditional, hardware-centric models to a cloud-based, consumption-driven approach, NaaS allows businesses to scale networks on demand while optimizing cost and performance. This chapter explores the various **service models of NaaS**, each designed to meet different enterprise needs, ranging from basic connectivity to sophisticated network operations.

3.1 Connectivity-as-a-Service (CaaS)

Connectivity-as-a-Service (CaaS) is the most foundational offering within the NaaS spectrum. It provides enterprises with on-demand, scalable network connectivity without the need to invest in physical infrastructure. CaaS delivers connectivity solutions that include point-to-point connections, multipoint connections, and cloud interconnections.

- **Components**:

    o **Direct Cloud Connectivity**: Enterprises can securely connect to cloud service providers (e.g., AWS, Microsoft Azure, Google Cloud) via direct links that bypass the public internet, ensuring low latency and enhanced security.

    o **Data Center Interconnect**: CaaS enables high-speed, reliable connections between different data centers, offering redundancy and failover capabilities.

    o **Internet Access**: Enterprises can also procure managed internet services with customizable bandwidth and quality of service (QoS) features.

- **Benefits**:

    o **Scalability**: Organizations can scale connectivity up or down as needed, avoiding the overprovisioning that typically occurs with traditional network contracts.

    o **Cost-Efficiency**: Pay-as-you-go pricing models mean companies only pay for the bandwidth they consume, leading to predictable and controllable costs.

- **Ease of Deployment**: Connectivity can be provisioned quickly and dynamically, bypassing the lengthy procurement and setup processes typical of traditional hardware installations.

3.2 Virtual Network Functions (VNF) as a Service

Virtual Network Functions (VNF) allow enterprises to deploy network services such as firewalls, load balancers, and VPNs in a virtualized environment. VNFs are a key part of the NaaS offering because they replace traditional network appliances with software-based functions, providing flexibility, agility, and reduced capital expenditure (CapEx).

- **Components:**

  - **Firewalls**: On-demand virtual firewalls ensure that enterprises can deploy and configure network security policies quickly.

  - **Load Balancers**: Virtualized load balancers optimize traffic distribution across servers and applications, enhancing performance and reliability.

  - **WAN Optimization**: VNFs can also optimize WAN performance by compressing data, deduplicating traffic, and improving latency.

- **Benefits**:

  - **Flexibility**: Enterprises can customize their virtual networks to specific business needs without being tied to fixed hardware configurations.

  - **Cost Savings**: VNFs eliminate the need for physical appliances, reducing hardware costs and simplifying maintenance.

  - **Rapid Deployment**: Virtual network functions can be deployed and reconfigured in minutes, enhancing responsiveness to changing business requirements.

3.3 Managed NaaS

Managed NaaS involves outsourcing the design, deployment, and management of an enterprise's network infrastructure to a NaaS provider. This model is particularly appealing for businesses that do not have extensive in-house IT resources or want to focus on core business functions rather than network management.

- **Components**:

    - **Network Monitoring and Management**: Managed NaaS providers oversee the entire network infrastructure, ensuring uptime, security, and performance. This includes proactive network health monitoring, troubleshooting, and updates.

    - **Security Management**: Providers often include security services such as intrusion detection, anti-DDoS protection, and secure access management as part of the managed service.

    - **Network Analytics**: Advanced analytics tools allow enterprises to track performance, usage patterns, and potential issues in real time, leading to informed decision-making and more efficient operations.

- **Benefits**:

    - **Reduced Operational Overhead**: Enterprises can offload the day-to-day management of their networks to specialized NaaS providers, freeing up internal resources.

- **Expertise**: Managed NaaS providers bring specialized knowledge and best practices, ensuring that networks are optimized for performance, security, and compliance.

  - **Predictable Costs**: Managed services typically operate on subscription-based models, offering predictable monthly costs with no unexpected capital expenditures.

3.4 Security-as-a-Service (SECaaS)

With the increasing complexity and frequency of cybersecurity threats, **Security-as-a-Service (SECaaS)** has emerged as a critical NaaS offering. This service delivers scalable, on-demand security functions such as firewalls, threat detection, and intrusion prevention, all managed through a cloud-based platform.

- **Components:**

  - **Intrusion Detection/Prevention Systems (IDS/IPS)**: These systems monitor traffic for malicious activity and block potential threats in real-time.

  - **DDoS Protection**: SECaaS provides enterprises with automated defenses against Distributed

Denial of Service (DDoS) attacks, which can cripple networks.

- **Endpoint Security**: SECaaS extends security to the enterprise's devices and endpoints, ensuring compliance with security policies and mitigating the risk of data breaches.

- **Benefits**:

    - **Scalable Security**: Enterprises can scale security services according to the growth of their networks and IT environments, paying only for what they need.

    - **Real-Time Threat Management**: Cloud-based security tools provide continuous monitoring and fast responses to potential threats, reducing risk exposure.

    - **Compliance**: SECaaS providers often have built-in compliance frameworks to meet industry regulations, simplifying the compliance process for enterprises.

3.5 Network Automation and Orchestration

Network Automation and **Orchestration** are evolving as crucial services within the NaaS ecosystem, allowing enterprises to automate routine network tasks and orchestrate complex, multi-cloud network environments.

- **Components:**

    - **Policy-Driven Automation**: Enterprises can set predefined network policies that automatically trigger changes in traffic routing, bandwidth allocation, and security configurations.

    - **End-to-End Orchestration**: NaaS platforms can orchestrate connectivity across multiple cloud environments and data centers, ensuring consistent network performance.

    - **Self-Healing Networks**: Advanced automation tools enable networks to detect and correct issues (e.g., rerouting traffic in case of a link failure) without human intervention.

- **Benefits:**

    - **Operational Efficiency**: Automation reduces manual intervention in routine tasks such as bandwidth management or security updates, freeing up IT resources.

- o **Improved Performance**: Automated networks can quickly adapt to changing traffic patterns, ensuring optimal performance under dynamic conditions.

- o **Faster Response to Issues**: Automated incident detection and resolution reduce downtime and mitigate the impact of network outages.

3.5 Conclusion

NaaS service models are reshaping how businesses approach their networking needs by offering on-demand, scalable, and flexible solutions that cover a wide range of functions—from basic connectivity to advanced security and network orchestration. Whether through Connectivity-as-a-Service, Virtual Network Functions, Managed NaaS, Security-as-a-Service, or Automation, these service models allow enterprises to offload the complexity of network management while enjoying greater agility, cost efficiency, and performance. As NaaS continues to evolve, these models will play a critical role in supporting enterprises' digital transformation efforts and adapting to the future of networking.

Chapter 4:

Implementing NaaS in Enterprises

Network-as-a-Service (NaaS) is transforming how enterprises manage their networks by providing flexibility, scalability, and simplified connectivity solutions. This chapter delves into the practical steps, benefits, and challenges of implementing NaaS in an enterprise environment. We will explore how businesses can leverage NaaS to optimize performance, improve cost efficiency, and achieve seamless cloud connectivity.

4.1 Understanding NaaS in Enterprise Context

NaaS enables enterprises to procure, scale, and manage network resources on-demand, like how cloud services are consumed. It abstracts traditional networking hardware into a service layer, allowing businesses to purchase network capabilities like bandwidth, VPNs, security services, and cloud interconnects as a utility.

Key NaaS features include:

On-demand scalability: Enterprises can increase or decrease network resources instantly based on current needs, without purchasing or maintaining physical hardware.

Cost efficiency: By moving from CapEx (capital expenditures) to OpEx (operational expenses), companies only pay for the resources they use, reducing costs associated with over-provisioning.

Automation and orchestration: NaaS platforms often provide automated network provisioning and management tools, which simplify day-to-day operations and improve agility.

4.2 Key Components of NaaS for Enterprises

When implementing NaaS, enterprises generally consider the following key components:

4.2.1 Virtual Routers

Virtual routers replace traditional physical routers, providing similar functionality without the need for dedicated hardware. These routers enable dynamic routing between multiple network endpoints and simplify network management. By leveraging virtual routers, businesses can manage multiple cloud connections with ease.

4.2.2 Cloud Connectivity

NaaS plays a critical role in facilitating direct connections to public clouds, ensuring low-latency and secure data transfers between on-premises infrastructure and cloud providers like AWS, Microsoft Azure, and Google Cloud. Many NaaS solutions offer built-in support for multi-cloud and hybrid cloud architectures, allowing enterprises to move workloads seamlessly across environments.

4.2.4 Software-Defined WAN (SD-WAN)

NaaS integrates SD-WAN technologies, enabling businesses to optimize wide-area network (WAN) performance by routing traffic intelligently across multiple paths, such as internet, MPLS, or direct cloud connections. SD-WAN provides better control and visibility over network traffic, improving both performance and security.

4.2.4 Security as a Service

NaaS solutions often come with integrated security services, such as firewalls, encryption, and DDoS protection, ensuring that enterprises maintain robust security postures without managing separate hardware. Security services are dynamically provisioned and adjusted based on network needs.

4.3 Steps to Implement NaaS in an Enterprise

4.3.1 Assessing Current Network Infrastructure

Before transitioning to NaaS, enterprises need to audit their existing network infrastructure. This includes evaluating on-premises network devices, bandwidth consumption, latency issues, security requirements, and overall network performance.

Key considerations:

Which parts of the network can be virtualized?

How much capacity will be needed on-demand?

Which applications will benefit most from flexible network services?

4.3.2 Selecting a NaaS Provider

Choosing the right NaaS provider is critical to a successful implementation. Providers such as Megaport, Polarin, and Equinix offer varying degrees of integration with cloud platforms, geographical reach, and service levels. Key factors to consider include:

Cloud compatibility: Ensure the provider offers seamless integration with the enterprise's preferred cloud platforms.

Geographical coverage: For global enterprises, ensure that the provider's network has the reach to connect distant offices and data centers.

Security: Evaluate the provider's security offerings, such as end-to-end encryption and compliance certifications (e.g., ISO, SOC 2).

4.3.3 Defining Use Cases

Not all business processes or applications need the flexibility offered by NaaS. Enterprises should define clear use cases where NaaS delivers the greatest value:

Burst capacity: Scale network resources up or down during high-demand periods, such as product launches or seasonal spikes.

Disaster recovery: Ensure that remote office connectivity and cloud backup systems can be quickly restored in the event of an outage.

Multi-cloud environments: Simplify the management of multiple cloud providers by routing traffic between them through a NaaS platform.

4.3.4 Deployment and Integration

Once a provider is chosen and use cases are defined, the next step is to deploy the NaaS solution. This involves integrating the NaaS platform with existing network monitoring tools and cloud management platforms. In many cases, the provider offers API integration and automation tools to streamline the setup process.

Enterprises may also need to train their IT staff to manage and monitor NaaS deployments effectively.

4.4 Benefits of NaaS Implementation

The implementation of NaaS delivers a range of business benefits:

Cost Savings: Moving to a pay-as-you-go model ensures enterprises only pay for what they use, eliminating the need to invest in expensive hardware upfront.

Improved Agility: With on-demand network provisioning, enterprises can respond to business needs faster, launching new services or scaling existing ones in minutes rather than months.

Simplified Management: Centralized control through a single NaaS platform reduces the complexity of managing multi-cloud and hybrid cloud networks, cutting down on manual configurations.

Enhanced Security: Integrated security features ensure data is protected as it moves across networks, helping enterprises comply with regulations such as GDPR or industry-specific standards.

4.5 Challenges and Considerations

Despite the numerous benefits, enterprises may face challenges when adopting NaaS:

Complexity in Migration: Transitioning from a legacy network to a NaaS-based model can be complex, especially for enterprises with a large, geographically dispersed network.

Vendor Lock-in: Enterprises should be cautious about getting locked into a single NaaS provider, as this could reduce flexibility in the future.

Performance Issues: Depending on the provider's infrastructure, latency or bandwidth issues could arise, particularly in regions with limited NaaS provider coverage.

4.6 The Future of NaaS in Enterprises

As businesses continue to move toward cloud-first strategies, the demand for flexible, scalable networking will only increase. NaaS providers are expected to integrate more advanced features, such as AI-driven network optimization, edge computing capabilities, and zero-trust security models. Enterprises that adopt NaaS early stand to gain a competitive advantage by being able to innovate and scale faster than their competitors.

NaaS is evolving from being a cutting-edge technology to a mainstream enterprise solution. As more businesses embrace this model, it will transform the way companies approach networking, making it a key enabler of digital transformation initiatives.

This chapter covered the essential steps to implement NaaS, its benefits, and challenges, providing enterprises with a roadmap for modernizing their network infrastructure.

Chapter 5:

NaaS Security and Compliance

As Network-as-a-Service (NaaS) rapidly gains popularity, enterprises need to ensure that this shift in network management does not compromise security or violate compliance standards. The flexibility and scalability offered by NaaS solutions also introduce unique security challenges that must be addressed through robust design, governance, and policy implementation.

This chapter delves into the critical aspects of **NaaS security,** and the necessary steps organizations should take to meet **compliance standards** while harnessing the benefits of this service model.

5.1 Security Challenges in NaaS Adoption

NaaS fundamentally changes how networks are deployed and managed. While it offers numerous benefits, its cloud-centric nature also opens the door to several security risks:

5.1.1 Multi-Tenant Environments

NaaS platforms typically operate on shared, multi-tenant infrastructure. This means multiple customers share the same physical resources, albeit logically isolated. Though this architecture offers cost and scalability advantages, it introduces risks associated with **data leakage** or **misconfiguration** of tenant isolation, potentially leading to unauthorized access.

5.1.2 Network Visibility and Control

Traditional networks offer granular control over hardware and data paths. NaaS, being an abstracted service, can reduce **visibility** into underlying network operations, making it harder for enterprises to monitor traffic patterns or detect threats in real-time. Loss of control over traffic routing, inspection, and policy enforcement is a major concern for enterprises that rely heavily on secure network flows.

5.1.3 Compliance in Multi-Cloud and Hybrid Cloud Environments

NaaS integrates multiple environments—on-premises data centers, hybrid clouds, and multi-cloud setups—each of which may have its own compliance requirements. This decentralized model complicates efforts to meet regulatory standards for data residency, encryption, and auditing.

5.2 Key Security Features in NaaS

To address these challenges, NaaS platforms come equipped with various security features. Enterprises must assess the security capabilities of NaaS providers to ensure they align with organizational policies and regulatory frameworks.

5.2.1 Encryption and Data Protection

Encryption is foundational to securing data in transit across NaaS networks. Leading NaaS providers offer **end-to-end encryption** using standards such as **AES-256** and **TLS** to protect sensitive data traversing different cloud environments and public internet links. This ensures that data remains confidential even if intercepted.

NaaS providers also enable encryption of traffic between multi-cloud environments and hybrid deployments, ensuring consistent protection regardless of the underlying infrastructure.

5.2.2 Network Segmentation and Isolation

NaaS platforms use virtual segmentation techniques such as **Virtual LANs (VLANs)** or **Virtual Routing and Forwarding (VRF)** to ensure that customer traffic is isolated within multi-tenant environments. Proper segmentation prevents lateral movement of threats from one tenant's environment to another, limiting the scope of a potential breach.

5.2.3 Access Control and Authentication

NaaS platforms incorporate **multi-factor authentication (MFA)**, **role-based access control (RBAC)**, and **zero-trust architecture** to regulate who can access network resources. By implementing strict identity verification protocols, enterprises can control access to their network infrastructure, minimizing insider threats and external breaches.

Zero-trust ensures that every access request, whether internal or external, is thoroughly authenticated and authorized, preventing unauthorized users from accessing critical resources.

5.2.4 Intrusion Detection and Prevention

NaaS platforms typically offer built-in **intrusion detection systems (IDS)** and **intrusion prevention systems (IPS)** to monitor network traffic for suspicious activity. These tools can automatically detect patterns associated with malware, distributed denial of service (DDoS) attacks, or other forms of cyberattacks, and take preemptive actions to block or mitigate these threats.

5.2.5 Network Monitoring and Threat Intelligence

With cloud-native network monitoring tools, NaaS platforms provide enterprises with **real-time visibility** into traffic flows, performance metrics, and security events. Integration with **security information and event management (SIEM)** platforms enables continuous threat monitoring and advanced analytics to identify potential vulnerabilities or anomalies.

5.3 Compliance Considerations in NaaS

Many industries are subject to regulatory frameworks governing how sensitive data is handled, stored, and transmitted. As enterprises migrate their networking infrastructure to a NaaS model, they must remain compliant with these standards to avoid penalties and legal consequences.

5.3.1 Data Privacy and Residency Requirements

NaaS solutions often operate across multiple jurisdictions, each with its own rules regarding **data residency** and **data privacy**. For example, regulations like the **General Data Protection Regulation (GDPR)** in Europe or **India's Data Protection Bill** mandate that sensitive data must remain within a particular geographic boundary.

When using NaaS, enterprises must ensure that their providers offer controls to specify where data is processed and stored, and whether it can cross international boundaries.

5.3.2 Industry-Specific Compliance

Different industries have specific compliance standards that may influence how NaaS services are configured:

- **Healthcare**: Organizations must adhere to **HIPAA** regulations, which dictate how protected health information (PHI) is transmitted and secured.

- **Finance**: Financial institutions need to comply with **PCI-DSS** (for payment data) and **Sarbanes-Oxley (SOX)** for financial reporting controls. NaaS providers must demonstrate that they can meet these rigorous security and audit requirements.

- **Government and Defense**: Compliance with frameworks such as **FISMA** and **FedRAMP** is mandatory for public sector organizations, requiring stringent controls over network security, encryption, and identity management.

5.3.3 Auditing and Reporting

Regulatory frameworks often require detailed auditing of network activity, including the ability to trace user actions and monitor changes in network configurations. NaaS platforms simplify compliance by offering built-in **audit logs, compliance reporting tools**, and **automated documentation**.

For instance, NaaS providers typically provide audit logs of every network change, user access request, or configuration modification to ensure traceability and accountability.

5.4 Best Practices for Securing NaaS Deployments

5.4.1 Vendor Due Diligence

Enterprises should thoroughly evaluate the **security posture** of NaaS providers before committing to a solution. This includes understanding the provider's security policies, infrastructure resilience, and compliance certifications (e.g., **ISO 27001, SOC 2,** or **FedRAMP**).

5.4.2 Integrating Security into Network Design

When deploying NaaS, enterprises must integrate security from the outset. This includes defining clear access control policies, segmenting network traffic, and implementing **network security controls** such as firewalls, IDS/IPS, and monitoring systems.

5.4.3 Continuous Monitoring and Updates

NaaS platforms often receive security updates and patches automatically. However, enterprises should adopt a policy of continuous monitoring and auditing to ensure their network configurations align with both organizational security policies and regulatory requirements.

5.4.4 Disaster Recovery and Business Continuity Planning

As with any cloud-based service, enterprises must establish robust **disaster recovery** and **business continuity plans** for NaaS. This ensures that, in the event of a cyberattack, network failure, or data loss, critical systems can be quickly restored.

5.5 The Future of NaaS Security

As enterprises continue to adopt NaaS and integrate with **5G, edge computing,** and **IoT networks,** the complexity of securing these services will increase. However, advancements in **AI-driven security, machine learning** for threat detection, and **zero-trust architectures** will enhance the ability of NaaS providers to address future threats.

NaaS vendors will also play a crucial role in helping enterprises navigate an increasingly complex compliance landscape by offering more automated tools for auditing, monitoring, and governance.

5.6 Conclusion

Security and compliance are paramount concerns for any organization adopting NaaS. By leveraging built-in security features such as encryption, segmentation, and access controls, and by adhering to industry-specific compliance standards, enterprises can confidently deploy NaaS while maintaining the integrity of their network infrastructure. Continuous monitoring, vendor evaluation, and the integration of security into network design will ensure a resilient, secure, and compliant network environment for enterprises moving forward.

Chapter 6:

Building a Business Case for NaaS

As enterprises increasingly consider Network-as-a-Service (NaaS) to modernize and streamline their network infrastructure, building a robust business case is essential to secure organizational buy-in. A well-crafted business case articulates not only the technical benefits of NaaS but also quantifies its economic, operational, and strategic impacts on the organization. This chapter explores the key components of building a compelling business case for NaaS adoption, with a focus on demonstrating **Return on Investment (ROI), Total Cost of Ownership (TCO)** savings, and aligning NaaS with broader business objectives.

6.1 Identifying the Strategic Need for NaaS

The first step in building a business case for NaaS is to identify the **strategic drivers** that make it a necessity for the organization. These drivers often revolve around the need for:

- **Scalability and Flexibility**: NaaS allows businesses to scale their networking capacity in real time to match fluctuating business demands, making it ideal for enterprises experiencing rapid growth or unpredictable demand patterns.

- **Cost Efficiency**: Unlike traditional network models, which require significant capital investments, NaaS offers a **subscription-based** or **pay-as-you-go** pricing model, reducing upfront capital expenditure (CapEx) and offering predictable operating expenditure (OpEx).

- **Cloud Integration**: As more organizations adopt **multi-cloud** and **hybrid cloud** strategies, the need for seamless, secure cloud interconnectivity becomes critical. NaaS supports this through flexible and secure cloud connections, minimizing complexity and improving performance.

- **Agility and Innovation**: By reducing time-to-market for new network services and eliminating the need for physical infrastructure upgrades, NaaS enables enterprises to respond quickly to market changes and drive innovation.

6.2 Assessing Business Impact

Once the strategic need is established, the next step is to assess how NaaS impacts the broader business. Focus on three key areas:

6.2.1 Operational Efficiency

Traditional network models often require specialized IT staff to manage and maintain physical network infrastructure. NaaS automates many of these tasks, reducing operational overhead and freeing up IT teams to focus on higher-value activities.

Key Points:

- **Automation**: NaaS automates network provisioning, scaling, and management, leading to fewer manual errors and faster deployments.

- **Reduced IT Maintenance**: By outsourcing network management to the NaaS provider, enterprises can decrease the need for in-house network expertise, leading to lower labor costs.

6.2.2 Business Agility

NaaS provides the flexibility to quickly adapt to changing business conditions, making it easier for enterprises to enter new markets, deploy new applications, or support remote workforces. The ability to **provision network resources** on-demand accelerates time-to-market, giving organizations a competitive edge.

Key Points:

- **On-Demand Scaling**: NaaS allows for real-time adjustments to network capacity, eliminating the need for lengthy procurement cycles associated with traditional networks.

- **Support for Innovation**: By leveraging NaaS, businesses can experiment with new applications, test new markets, and support digital transformation initiatives without being constrained by network limitations.

6.2.3 Risk Mitigation

Security, reliability, and compliance are essential for business continuity. NaaS providers typically offer **built-in security features**, **Service Level Agreements (SLAs)**, and **compliance tools** to mitigate risks related to network outages, cyber threats, and regulatory requirements.

Key Points:

- **Improved Network Uptime**: NaaS providers often guarantee **99.99% availability** or higher, reducing the risk of downtime that could impact business operations.

- **Security Integration**: NaaS platforms provide advanced security features such as encryption, access control, and threat detection, reducing the risk of data breaches and cyberattacks.

6.3 Economic Benefits: TCO

A central component of any business case is demonstrating the **economic value** of NaaS adoption through a detailed TCO analysis.

6.3.1 Total Cost of Ownership (TCO) Comparison

Traditional network models require significant capital investment in physical hardware (routers, switches, cables, etc.), as well as ongoing operational costs for maintenance, upgrades, and energy consumption. NaaS, by contrast, eliminates the need for hardware purchases and offers a **subscription-based pricing model** that consolidates costs into a predictable OpEx.

Key TCO Components of Traditional Networks:

- **Hardware acquisition costs**: Initial capital investment in physical infrastructure.

- **Maintenance and support**: Ongoing maintenance contracts, repairs, and troubleshooting.

- **Energy and cooling**: Power consumption for running and cooling data center hardware.

- **Labor costs**: In-house IT staff required for day-to-day network management and troubleshooting.

Key TCO Components of NaaS:

- **Subscription fees**: Predictable, recurring costs based on usage (bandwidth, number of connections, etc.).

- **Reduced labor costs**: Minimal IT staff needed for network management.

- **No hardware costs**: NaaS is fully managed by the provider, removing hardware-related expenses.

- **Scalability**: Costs scale with usage, preventing overprovisioning or underutilization of resources.

TCO Reduction Example:

Category	Traditional Network (Yearly)	NaaS (Yearly)
Hardware Costs	$500,000	$0
Maintenance Contracts	$100,000	$10,000
IT Staff Costs	$150,000	$50,000

Energy/Cooling Costs	$50,000	$0
Total Costs	$800,000	$60,000

In this example, transitioning to NaaS leads to a significant reduction in overall TCO, with NaaS costing **25% or less** of traditional network models.

6.4 Risk Management and Mitigation

A comprehensive business case should also address the **risks** associated with NaaS adoption, as well as the mitigation strategies in place to minimize them. These risks include:

- **Vendor Lock-In**: Enterprises need to ensure that NaaS providers offer interoperability and easy migration paths to prevent long-term dependency on a single vendor.

- **Security Concerns**: As discussed in the previous chapter, robust encryption, access control, and monitoring solutions must be in place to protect enterprise data.

- **Service Interruptions**: Enterprises should select NaaS providers with strong SLAs and redundancy protocols to mitigate the risk of downtime.

Mitigation strategies can involve conducting thorough vendor assessments, implementing multi-cloud strategies to avoid lock-in, and continuously monitoring network performance through integrated security and performance tools.

6.5 Aligning NaaS with Business Objectives

Finally, the business case should clearly align NaaS adoption with broader **business objectives**. These include:

- **Digital Transformation**: NaaS supports digital transformation by offering flexible, scalable, and cloud-native networking solutions that align with modern IT infrastructure.

- **Customer Experience**: Improved network performance, reliability, and uptime contribute to a better user experience for both employees and customers, driving greater customer satisfaction and retention.

- **Competitive Advantage**: Enterprises that can scale and innovate faster through NaaS are better positioned to outpace competitors in the digital marketplace.

6.6 Conclusion

Building a business case for NaaS is not solely about reducing costs—it's about demonstrating how NaaS aligns with the enterprise's strategic goals of agility, innovation, and efficiency.

By focusing on the financial benefits (TCO and ROI), operational improvements, and risk mitigation strategies, enterprises can make a compelling case for why NaaS is not just a network upgrade, but a transformative investment for the future.

Chapter 7:

Economic Implications of NaaS

The shift toward **Network-as-a-Service (NaaS)** is reshaping how enterprises approach networking, with far-reaching economic implications. NaaS represents a new economic model for network infrastructure, enabling organizations to purchase and scale network resources on demand. In this chapter, we explore the direct and indirect economic effects of adopting NaaS across industries, focusing on operational efficiencies, cost structures, and macroeconomic impacts.

7.1 Shift from CapEx to OpEx

One of the most significant economic implications of NaaS is the transformation from **capital expenditures (CapEx)** to **operational expenditures (OpEx)**. Traditional networking models require enterprises to make large upfront investments in physical infrastructure—routers, switches, and data center hardware. With NaaS, these costs are replaced with a subscription-based or pay-as-you-go model, allowing enterprises to convert fixed costs into variable ones.

Benefits of OpEx:

- **Lower Upfront Costs**: Enterprises no longer need to purchase expensive hardware. Instead, they pay for network services based on actual usage, reducing the financial barrier to entry.

- **Improved Cash Flow**: The shift to an OpEx model helps businesses preserve cash flow by spreading costs over time, rather than requiring a large initial outlay.

- **Scalability**: Businesses can scale their networking infrastructure up or down as needed, avoiding over-investment and under-utilization.

From an economic perspective, this shift drives greater financial agility, particularly for small and medium-sized enterprises (SMEs) that may have been limited by the high costs of traditional network infrastructure.

7.2 Operational Efficiency and Productivity Gains

NaaS offers substantial productivity and operational efficiency benefits, which translate into economic gains:

7.2.1 Automation and Simplified Management

With NaaS, much of the network management is automated. This reduces the need for manual configurations and on-site maintenance, leading to lower labor costs and reduced downtime. Enterprises can provision network resources within minutes through a cloud-based platform, significantly improving operational agility.

7.2.2 Outsourced Expertise

NaaS providers often bundle their services with **network management**, **security**, and **monitoring**, enabling enterprises to offload complex network operations to specialized providers. This reduces the need for in-house expertise, lowering training and staffing costs.

These efficiency gains contribute to increased productivity, as IT teams can focus on higher-value tasks rather than routine network management. Furthermore, by reducing downtime, NaaS helps avoid potential economic losses related to network outages.

7.3 Cost Savings and Efficiency in Network Usage

The **pay-as-you-go (PAYG)** model central to NaaS helps enterprises optimize network costs. Companies no longer need to over-provision for peak usage or worry about under-utilizing expensive network equipment. This model allows businesses to pay only for what they use, leading to better cost control and predictability.

7.3.1 Cost Predictability

Traditional network costs are often difficult to forecast, as they are influenced by equipment life cycles, maintenance fees, and unpredictable scaling needs. With NaaS, enterprises enjoy **cost transparency**, with predictable billing models that tie directly to usage.

7.3.2 Lower Total Cost of Ownership (TCO)

By eliminating the need for physical infrastructure, NaaS reduces **hardware maintenance, power consumption**, and **real estate costs** (e.g., housing equipment in data centers). This translates into a significantly lower total cost of ownership over the long term.

According to an IDC study, companies adopting NaaS can achieve **20-70% cost savings** in their networking budgets due to reduced hardware expenses, fewer maintenance costs, and more efficient network utilization.

7.4 Economic Impact of Cloud and Multi-Cloud Connectivity

NaaS plays a crucial role in supporting **cloud-first strategies** and enabling **multi-cloud** environments. The economic implications of enhanced cloud connectivity extend beyond networking cost savings:

7.4.1 Acceleration of Cloud Adoption

With streamlined connectivity to public cloud platforms like **AWS, Azure,** and **Google Cloud**, NaaS reduces the complexity of managing cloud-based applications. This accelerates cloud adoption by making it easier for enterprises to move workloads and data across different cloud providers.

7.4.2 Increased Cloud ROI

By facilitating seamless connectivity between cloud platforms and on-premises infrastructure, NaaS enhances cloud performance and optimizes the use of cloud services. This improves the **return on investment (ROI)** of cloud projects, as businesses can better leverage cloud resources without worrying about network constraints.

7.5 Macroeconomic Impacts of NaaS Adoption

At the macroeconomic level, widespread NaaS adoption has the potential to influence national and global economies in several ways:

7.5.1 Enabling Digital Transformation

NaaS is a critical enabler of digital transformation across industries. As businesses become more connected, they can harness the power of data, automation, and cloud computing to create new products and services. This digital transformation drives innovation and contributes to the growth of new markets.

7.5.2 Impact on the IT and Telecommunications Industries

As enterprises transition to NaaS, demand for traditional networking equipment may decline. This could have a negative impact on hardware manufacturers, while benefiting **software-defined networking (SDN)** providers, cloud service providers, and NaaS vendors. At the same time, telecommunications companies may face increased pressure to adopt new business models to compete with cloud-based NaaS offerings.

7.5.3 Job Market Shifts

The rise of NaaS will likely lead to changes in the IT job market. The demand for network engineers and hardware technicians may decrease, while roles in **cloud architecture, network automation,** and **cybersecurity** will become increasingly important. Enterprises may need to invest in retraining their workforce to meet the evolving demands of the NaaS-driven economy.

7.6 Challenges and Risks

While NaaS offers numerous economic advantages, it is not without challenges:

7.6.1 Vendor Lock-In

Enterprises relying heavily on a single NaaS provider may face **vendor lock-in,** limiting their flexibility to switch providers without incurring high migration costs. This can negate some of the cost-saving benefits of NaaS if businesses are tied to unfavourable contracts.

7.6.2 Data Security and Compliance Costs

NaaS providers handle sensitive network data, raising concerns about security and compliance. Enterprises may need to invest in additional security measures or audits to ensure data protection, particularly in industries with stringent regulatory requirements (e.g., healthcare and finance).

The economic benefits of Network-as-a-Service (NaaS) can be analysed using **Total Cost of Ownership (TCO)** and **Return on Investment (ROI)** calculations. These metrics help enterprises evaluate the cost-effectiveness of transitioning from traditional networking infrastructure to NaaS. In this section, we explore these financial metrics to highlight the long-term savings and potential returns from NaaS adoption.

7.7 Total Cost of Ownership (TCO) Analysis for NaaS

TCO measures the complete cost of owning a product or service over its entire lifecycle, including both direct and indirect costs. For NaaS, TCO primarily includes the following components:

- **Subscription Fees or Pay-as-You-Go Costs**: The core expense of NaaS services, which varies based on usage (bandwidth, VPNs, cloud interconnects, etc.).

- **Operational Costs**: These include management and maintenance, which are significantly lower with NaaS compared to traditional networks because much of the infrastructure is virtualized and automated.

- **Labor Costs**: IT staff time and expertise required to maintain physical hardware are drastically reduced with NaaS, as network management and troubleshooting are often outsourced to the NaaS provider.

- **Energy and Cooling Costs**: On-premises physical networking equipment consumes energy, requiring cooling systems, which adds to TCO. NaaS reduces or eliminates this overhead.

- **Hardware Depreciation**: Traditional network infrastructure has a finite lifespan and loses value over time, whereas NaaS allows enterprises to avoid capital depreciation entirely.

TCO Breakdown Example for Traditional Networking vs. NaaS

Cost Category	Traditional Network (Annual)	NaaS (Annual)
Hardware Costs	$200,000	$0 (no physical assets)
Maintenance/Support Contracts	$30,000	$5,000
Labor Costs	$120,000 (IT staff for network)	$70,000 (reduced staff)
Energy and Cooling	$15,000	$0
Network Subscription/Usage	N/A	$150,000 (NaaS fees)
Total Annual Cost	$365,000	$195,000

In this example, transitioning to NaaS results in **77% TCO savings** compared to traditional networking infrastructure, primarily driven by reduced hardware, energy, and labor costs.

7.8 ROI Calculation for NaaS Projects

ROI is a critical metric for evaluating the financial returns of NaaS implementation. It compares the net gains (or savings) from NaaS adoption to the initial investment required, giving enterprises insight into the value of their expenditure.

The ROI formula is:

ROI= (Net Gains (Savings) from NaaS/Initial Investment) ×100

Key Elements of ROI Calculation for NaaS:

1. **Initial Investment**: For NaaS, this includes any upfront setup fees, consulting costs, or integration expenses, which are generally much lower than for traditional networks.

2. **Net Gains (Savings)**: These are calculated by subtracting the total cost of NaaS (over a given period) from the traditional network TCO over the same period. Savings may include reductions in hardware, maintenance, staffing, and energy costs, as well as revenue from increased agility, faster deployments, and network uptime.

ROI Example Calculation

Let's assume an enterprise is evaluating a 3-year NaaS implementation project. The traditional network setup has a TCO of **$365,000 per year**, while NaaS costs **$195,000 per year**. Initial integration and setup costs for NaaS are **$50,000**.

- **Total TCO for Traditional Network (over 3 years):**
  $365,000 × 3 = **$1,095,000**

- **Total TCO for NaaS (over 3 years):**
  ($195,000 × 3) + $50,000 (setup costs) = **$635,000**

- **Net Gains (Savings) from NaaS:**
  $1,095,000 – $635,000 = **$760,000**

Now, the ROI calculation:

$$ROI = (760{,}000/50{,}000) \times 100 = 920\%$$

In this example, the NaaS project delivers a **920% ROI** over 3 years, demonstrating the significant financial returns of transitioning to a NaaS model.

7.9 Additional Benefits Impacting TCO and ROI

7.9.1 Network Agility and Time-to-Market

NaaS provides enterprises with rapid provisioning and dynamic scalability, which directly impacts business agility. Enterprises can quickly expand or reduce network capacity based on demand, enabling faster rollouts of new services. This **reduced time-to-market** for new products and services leads to increased revenue generation, indirectly improving ROI.

7.9.2 Enhanced Network Uptime

NaaS providers offer service level agreements (SLAs) with **99.99% uptime guarantees** or higher. Reduced downtime translates into lower revenue losses from outages and improves business continuity. Improved uptime increases productivity and reduces costs associated with network failures, which boosts the overall ROI.

7.9.3 Reduced Risk and Compliance Costs

NaaS vendors typically include integrated **security** and **compliance** features, such as encryption, firewalls, and monitoring. By reducing the need for enterprises to manage these aspects in-house, NaaS lowers compliance risks and mitigates the costs of data breaches or regulatory fines, further enhancing ROI.

7.10 Long-Term Economic Benefits of NaaS Adoption

NaaS offers long-term financial advantages that extend beyond initial savings:

- **Flexible Growth**: As enterprises grow, they can easily scale their networks without investing in new hardware, maintaining cost efficiency over time.

- **Ongoing Innovation**: NaaS providers regularly update and optimize their services, ensuring enterprises benefit from the latest technology without additional capital investment.

- **Global Reach**: NaaS allows enterprises to expand operations internationally without the complexity and cost of deploying physical infrastructure in new regions.

7.11 The Future of NaaS and Its Economic Potential

Looking ahead, the economic potential of NaaS is vast. As businesses continue to embrace cloud computing, edge computing, and IoT (Internet of Things), the demand for flexible, scalable networking solutions will grow. **AI-driven network management** and **5G adoption** will further enhance the value of NaaS, driving new use cases and efficiencies.

In addition, as NaaS providers expand their service offerings and geographical reach, they will open new markets for small and medium-sized enterprises (SMEs) that previously lacked access to advanced networking technologies.

7.12 Conclusion

The economic implications of NaaS adoption are profound, offering businesses a pathway to reduce costs, improve efficiency, and accelerate digital transformation. While challenges such as vendor lock-in and security concerns remain, the overall economic impact of NaaS is overwhelmingly positive, enabling enterprises to operate more flexibly and efficiently in an increasingly interconnected world.

Chapter 8:

Future Trends in NaaS

Network-as-a-Service (NaaS) is rapidly evolving, driven by increasing enterprise demands for agility, scalability, and cost efficiency in managing their network infrastructure. As businesses continue to embrace cloud-based solutions and the Internet of Things (IoT) proliferates, the NaaS market is set to experience transformative growth and technological advancements. This chapter explores key future trends in NaaS that will shape its trajectory over the coming years, enabling enterprises to capitalize on emerging opportunities and challenges in network architecture.

8.1 Edge Computing and NaaS Integration

As the need for low-latency, high-bandwidth services increases, especially with applications like autonomous vehicles, smart cities, and real-time analytics, **edge computing** is becoming more critical. NaaS providers are increasingly integrating edge computing capabilities into their offerings, enabling data processing closer to the source of data generation.

- **Trend**: NaaS platforms will increasingly offer seamless integration with edge data centers, allowing enterprises to optimize their workloads for low-latency performance at the edge. This will support mission-critical applications, especially those requiring real-time analytics and decision-making.

- **Impact**: The combination of NaaS with edge computing will enable enterprises to extend their networks closer to end-users and devices, reducing latency and improving performance for applications like IoT, augmented reality (AR), and virtual reality (VR). This also leads to enhanced customer experiences and more reliable service delivery.

8.2 AI and Automation in NaaS

Artificial Intelligence (AI) and automation are playing a transformative role in optimizing network management and operations. With increasing complexity in managing multi-cloud, hybrid environments, and large-scale IoT deployments, AI-driven NaaS is poised to streamline network administration and decision-making.

- **Trend**: AI will drive the **next wave of automation in NaaS**, enabling self-configuring, self-healing, and self-optimizing networks. AI-powered tools will predict network failures, optimize bandwidth allocation, and enforce security policies automatically.

- **Impact**: AI and machine learning (ML) algorithms can continuously monitor network performance and usage patterns to offer predictive maintenance, reducing downtime and improving network resilience. Enterprises will benefit from reduced operational overhead and more agile, responsive networks that can adapt to changing traffic conditions in real time.

8.3 Rise of 5G and NaaS

The rollout of **5G networks** is set to revolutionize the telecommunications industry, offering ultra-low latency, high bandwidth, and massive device connectivity. This next-generation network infrastructure presents significant opportunities for NaaS providers to deliver enhanced connectivity services.

- **Trend**: 5G will create new possibilities for NaaS by offering ultra-fast, low-latency mobile connectivity, enabling seamless networking for IoT, smart cities, and edge applications. NaaS will also support private 5G networks for enterprises, allowing them to create customized, high-performance networks tailored to specific business needs.

- **Impact**: Enterprises can leverage 5G-enabled NaaS to deliver real-time services such as autonomous industrial operations, telemedicine, and connected logistics. This will also lead to the rise of **network slicing**, where NaaS providers will offer differentiated network services with guaranteed performance levels based on user requirements.

8.4 Hybrid Multi-Cloud Architectures

The shift toward **hybrid multi-cloud environments** is gaining momentum as enterprises recognize the need for flexibility and resilience in their network architectures. Hybrid multi-cloud allows businesses to distribute workloads across public clouds, private clouds, and on-premises data centres based on performance, security, and cost requirements.

- **Trend**: NaaS will continue to evolve as the backbone of hybrid multi-cloud strategies, enabling seamless interconnectivity between diverse cloud environments. This will allow enterprises to optimize their workloads, improve network efficiency, and reduce the complexity of managing different network architectures.

- **Impact**: By integrating multiple cloud environments, NaaS will allow enterprises to create **agile, dynamic networks** that can scale and adapt to workload shifts. This is especially critical for applications like disaster recovery, where workloads can be quickly moved between clouds to maintain continuity and compliance.

8.5 Software-Defined Networking (SDN) and NaaS

Software-defined networking (SDN) is a key enabler of NaaS, allowing for more flexible, programmable, and centralized control of network resources. SDN allows enterprises to decouple the control plane from the data plane, making network management more dynamic and efficient.

- **Trend**: SDN will continue to enhance the flexibility of NaaS offerings by allowing more granular control over network resources and policies. Enterprises will have more control over their networking infrastructure through APIs, enabling them to easily adjust network configurations, manage traffic flows, and optimize performance on-demand.

- **Impact**: The convergence of SDN and NaaS will offer enterprises greater control over their network architectures, allowing for more **dynamic and adaptive networking** that can respond to traffic spikes, security threats, or changes in workload demand. This will lead to increased cost savings and improved performance for enterprise networks.

8.6 Security-as-a-Service (SECaaS) and NaaS

As enterprises continue to adopt cloud-native architectures, security remains a top concern. **Security-as-a-Service (SECaaS)** is emerging as a key component of NaaS, offering integrated, cloud-native security services that address modern cybersecurity challenges.

- **Trend**: NaaS providers will increasingly bundle **SECaaS** with their connectivity offerings, allowing enterprises to easily enforce network security policies, implement encryption, and protect against DDoS attacks. This will simplify network security management, ensuring compliance and reducing the risk of data breaches.

- **Impact**: By integrating security at the network layer, NaaS can offer end-to-end protection for cloud-based and on-premise environments. This will be especially critical for enterprises managing sensitive data or operating in regulated industries where compliance with security standards like GDPR and HIPAA is mandatory.

8.7 Blockchain and NaaS

Blockchain technology, with its emphasis on decentralized, immutable record-keeping, is beginning to intersect with networking technologies like NaaS. Blockchain can offer greater transparency and security in managing network infrastructure and data transmission.

- **Trend**: NaaS platforms may incorporate **blockchain technology** to enhance network security, transparency, and trust. Blockchain could be used to manage and authenticate network transactions, maintain audit trails, and ensure data integrity across distributed networks.

- **Impact**: Blockchain can create more **secure, verifiable networks**, reducing the risk of tampering or fraud in network transactions. This would be particularly useful in industries such as finance, supply chain management, and healthcare, where data security and transparency are critical.

8.8 Sustainability and Green Networking

Environmental sustainability is becoming a key focus for enterprises, and NaaS can play a role in reducing the environmental impact of networking infrastructure.

- **Trend**: NaaS providers will adopt **green networking** practices, such as optimizing energy usage in data centers, reducing e-waste through virtualized networking hardware, and using AI to minimize power consumption.

- **Impact**: Enterprises can leverage NaaS to reduce their **carbon footprint**, align with sustainability goals, and meet regulatory requirements for environmental impact. Green NaaS offerings will appeal to organizations focused on corporate social responsibility (CSR) and long-term sustainability.

8.9 Conclusion

NaaS is at the forefront of a networking revolution, driven by advancements in AI, 5G, SDN, and edge computing. As these technologies mature, NaaS will become more agile, intelligent, and secure, helping enterprises build the next generation of network infrastructures. The future of NaaS lies in its ability to adapt to emerging trends while addressing the complexities of modern networking. Enterprises that embrace these trends early will be well-positioned to capitalize on the full potential of NaaS and build resilient, scalable networks that drive innovation and growth.

Chapter 9:

Industry Specific NaaS Applications

Network-as-a-Service (NaaS) has revolutionized the way industries manage their network infrastructure, providing tailored solutions that address specific needs and challenges across various sectors. This chapter explores the diverse applications of NaaS in key industries, highlighting how businesses can leverage these services to enhance operational efficiency, improve customer experiences, and drive innovation.

9.1 Healthcare

In the healthcare sector, where the timely sharing of critical information is essential, NaaS can significantly improve connectivity and data management.

- **Telemedicine**: NaaS enables secure, high-speed connections that facilitate telemedicine services. Healthcare providers can offer remote consultations, reducing patient wait times and improving access to care.

- **Data Sharing**: With NaaS, hospitals and clinics can securely share patient records and medical imaging across different locations. This enhances collaboration among healthcare professionals, leading to better patient outcomes.

- **IoT Integration**: Healthcare providers can leverage NaaS to connect medical devices and wearables that monitor patient health in real time. This data can be transmitted securely to cloud platforms for analysis, enabling proactive patient care.

9.2 Financial Services

The financial industry requires robust and secure networking solutions to protect sensitive customer data and ensure compliance with regulatory standards.

- **Secure Transactions**: NaaS provides the infrastructure necessary for secure online transactions, utilizing encrypted connections and dedicated network resources to minimize fraud risk.

- **Real-Time Analytics**: Financial institutions can deploy NaaS to connect to cloud-based analytics tools, allowing them to process and analyze vast amounts of data quickly. This enables better decision-making and enhances customer experiences through personalized services.

- **Disaster Recovery**: NaaS offers financial firms the ability to set up disaster recovery solutions quickly. With cloud-based backups and redundant network connections, businesses can ensure continuity in case of system failures or cyberattacks.

9.3 Retail

In the retail sector, NaaS plays a pivotal role in creating seamless shopping experiences and optimizing supply chain management.

- **Omnichannel Retailing**: NaaS facilitates connectivity between online and offline retail channels, allowing businesses to manage inventory in real time and provide customers with consistent experiences across platforms.

- **Point of Sale (POS) Systems**: Retailers can leverage NaaS to enhance their POS systems, ensuring fast and secure transactions regardless of location. This is especially crucial for seasonal or pop-up stores that may not have permanent infrastructure.

- **Supply Chain Optimization**: NaaS can connect various stakeholders in the supply chain, enabling real-time communication and data sharing between suppliers, warehouses, and retailers. This leads to improved inventory management and reduced delays.

9.4 Manufacturing

The manufacturing sector benefits significantly from NaaS through improved connectivity and automation capabilities.

- **IoT and Industry 4.0**: NaaS supports the integration of IoT devices in manufacturing, allowing for real-time monitoring of production lines. This data can be analyzed to optimize processes, predict maintenance needs, and enhance product quality.

- **Supply Chain Collaboration**: Manufacturers can use NaaS to collaborate with suppliers and distributors seamlessly. Secure connections enable data sharing regarding inventory levels, order statuses, and production schedules, fostering greater supply chain efficiency.

- **Remote Monitoring and Control**: NaaS allows manufacturers to monitor and control machinery remotely, reducing the need for on-site personnel and enabling quicker responses to production issues.

9.5 Education

In the education sector, NaaS enhances connectivity for institutions, educators, and students, facilitating improved learning environments.

- **Remote Learning**: Educational institutions can leverage NaaS to provide secure and reliable connectivity for online courses and virtual classrooms, ensuring that students have access to high-quality education regardless of their location.

- **Collaborative Tools**: NaaS enables the deployment of collaborative platforms that allow students and educators to work together in real time. This can enhance project-based learning and foster greater engagement among students.

- **Administrative Efficiency**: NaaS can streamline administrative processes by connecting various systems, such as student information systems, financial aid management, and learning management systems, improving overall operational efficiency.

9.6 Transportation and Logistics

The transportation and logistics industry relies heavily on real-time data and communication, making NaaS essential for optimizing operations.

- **Fleet Management**: NaaS allows for seamless connectivity between vehicles and central management systems, enabling real-time tracking of shipments, route optimization, and improved fleet utilization.

- **Smart Warehousing**: In logistics, NaaS can connect IoT devices used in warehouses, providing real-time inventory management and automation of storage and retrieval processes.

- **Customer Communication**: NaaS facilitates communication between logistics providers and customers, allowing for real-time updates on shipment status and improving customer satisfaction.

9.7 Government and Public Sector

Governments and public sector organizations can leverage NaaS to enhance service delivery and improve citizen engagement.

- **Public Safety Networks**: NaaS enables secure, reliable communication networks for emergency services, ensuring that first responders can access critical information in real time.

- **Smart City Initiatives**: Governments can use NaaS to support smart city projects that integrate IoT devices for traffic management, public transportation, and environmental monitoring, enhancing the quality of urban life.

- **Citizen Services**: NaaS facilitates online services for citizens, such as tax filing, permit applications, and access to public records, improving the efficiency of government operations and citizen engagement.

9.8 Conclusion

The applications of NaaS across various industries illustrate its versatility and capability to address specific business needs. By providing tailored connectivity solutions, NaaS enables organizations to enhance operational efficiency, improve customer experiences, and foster innovation. As businesses continue to adopt NaaS, the potential for industry-specific applications will only expand, creating new opportunities for growth and transformation in the digital age. Through this industry-focused lens, NaaS is not just a technological advancement but a critical enabler of strategic initiatives across diverse sectors.

Chapter 10:

NaaS Challenges and Limitations

While Network-as-a-Service (NaaS) offers compelling benefits such as scalability, flexibility, and cost efficiency, it is not without challenges and limitations. Enterprises considering adopting NaaS must understand these potential barriers to fully leverage the technology and make informed decisions about its integration into their existing infrastructure. This chapter outlines the key challenges enterprises face when deploying NaaS and offers insights into overcoming these limitations.

10.1 Dependency on Third-Party Providers

One of the primary challenges of adopting NaaS is the dependency on third-party providers for critical network infrastructure. By outsourcing network management to a NaaS provider, enterprises cede a certain degree of control over their network operations, which can lead to:

- **Vendor Lock-In**: Once a company has invested heavily in a particular NaaS provider, migrating to another provider may be complex, time-consuming, and costly. Different vendors may use proprietary protocols, APIs, or configurations, making seamless migration difficult. This lack of interoperability can limit the organization's flexibility in choosing or switching providers as business needs evolve.

- **Limited Customization**: Some NaaS platforms may offer limited customization options compared to traditional on-premises network setups. While NaaS is built to be scalable and flexible, businesses with unique network configurations may find it challenging to tailor solutions precisely to their needs, leading to potential performance inefficiencies.

Mitigation Strategies:

- Choose NaaS providers that offer **open standards** and **interoperability** to avoid long-term vendor lock-in.

- Opt for **multi-vendor strategies** to spread risk and avoid over-reliance on a single provider.

10.2 Security and Compliance Concerns

As with any cloud-based service, security remains a top concern when adopting NaaS. Enterprises must trust their NaaS providers to handle sensitive data securely and ensure network resilience against cyber threats. However, outsourcing network functions can introduce several security-related challenges:

- **Data Privacy and Compliance**: Depending on the industry, enterprises may have stringent regulatory requirements around data handling, such as GDPR or HIPAA. NaaS providers may store data across multiple locations, including different regions or countries, potentially leading to regulatory compliance issues if data is mishandled or stored in regions with less stringent data protection laws.

- **Shared Infrastructure Risks**: Since NaaS typically operates on a shared infrastructure model, there is an inherent risk that vulnerabilities in one tenant's network could impact others. While providers invest heavily in security, businesses must evaluate how well a NaaS provider can isolate and secure network resources for each client.

- **Visibility and Control**: With NaaS, enterprises may have less granular control over network security measures than they would with an on-premises solution. This reduced visibility into traffic patterns, logs, and security events can make it more difficult for IT teams to detect and respond to threats quickly.

Mitigation Strategies:

- Ensure that the NaaS provider complies with relevant **security certifications** (e.g., ISO 210001, SOC 2) and **data residency requirements**.

- Leverage **end-to-end encryption** and **robust access controls** to safeguard sensitive data.

- Use **third-party security monitoring tools** to gain deeper visibility into NaaS network activity and detect threats in real-time.

10.3 Performance Variability

NaaS delivers its services over the internet or other public networks, which can result in performance variability. Network congestion, latency, and packet loss are some of the issues enterprises may face when using NaaS, particularly for applications that require **high performance** or **low-latency connections**. The quality of service may vary based on factors such as:

- **Geographic Distance**: Long distances between an enterprise's location and the NaaS provider's data centers can lead to higher latency, which may affect real-time applications like video conferencing, VoIP, or online collaboration tools.

- **Network Congestion**: Depending on the provider's infrastructure and traffic patterns, enterprises may experience fluctuating performance, especially during peak usage periods or if the provider's backbone is under stress.

- **Service Level Agreements (SLAs)**: While most NaaS providers offer SLAs guaranteeing uptime and performance, the specifics of these agreements may not always align with an enterprise's expectations. Downtime, slow response times, or failures to meet SLAs can directly impact business operations.

Mitigation Strategies:

- Choose providers with strong SLAs and **geo-redundant data centers** to minimize the impact of geographic distance and network congestion.

- Consider **hybrid NaaS models**, where critical workloads can remain on-premises while non-mission-critical tasks are outsourced to NaaS providers.

10.4 Integration with Legacy Infrastructure

Many enterprises operate legacy systems and networks that are tightly integrated with their business processes. Shifting to a NaaS model may not always be straightforward, as it may require integrating modern NaaS platforms with older, often rigid, legacy infrastructure. Challenges include:

- **Complexity of Migration**: Migrating from traditional networking to NaaS may require significant reconfiguration of existing systems, redesigning workflows, and retraining IT staff. Businesses may need to continue operating in a hybrid model during the transition, which can introduce complexity in managing both legacy and cloud-based networks simultaneously.

- **Incompatible Protocols**: Some legacy systems may not support the advanced protocols and APIs that NaaS relies on. This can cause connectivity or compatibility issues that hinder seamless integration and require additional investment in upgrading existing hardware or software.

Mitigation Strategies:

- Take a **phased approach** to NaaS adoption, focusing on use cases or departments that are easier to transition first.

- Collaborate with NaaS providers who offer **customizable solutions** that can work alongside existing infrastructure during the transition phase.

10.5 Cost Management and Predictability

One of the key promises of NaaS is **cost savings**, thanks to its pay-as-you-go (PAYG) pricing model. However, while NaaS can reduce upfront capital expenditure (CapEx), enterprises may find it challenging to manage ongoing operating costs (OpEx), particularly if usage patterns are unpredictable. Challenges include:

- **Cost Overruns**: NaaS's flexible, usage-based pricing model can lead to unpredictable costs. If usage spikes due to unforeseen demand, enterprises might find themselves facing higher bills than anticipated. Additionally, as businesses scale, the total cost of running a NaaS solution may escalate more quickly than expected.

- **Hidden Costs**: Enterprises may face additional charges for features like advanced security, extra bandwidth, or premium support, which may not be included in base subscription plans.

Mitigation Strategies:

- Monitor NaaS usage closely and set **budgets or usage thresholds** to control costs.

- Use providers that offer **transparent pricing models** and detailed cost breakdowns to ensure predictability in monthly expenses.

10.6 Limited Industry-Specific Solutions

NaaS is a generalized service that may not always offer tailored solutions for industry-specific requirements. Certain sectors, such as healthcare or finance, have unique networking needs related to compliance, security, and data management. Off-the-shelf NaaS offerings may not fully align with these specific needs, making it harder for enterprises in highly regulated industries to adopt the service.

Mitigation Strategies:

- Work with NaaS providers who offer **industry-specific solutions** or allow for extensive customization.

- Partner with third-party consultants or system integrators who can help customize NaaS offerings to fit the unique needs of specific sectors.

10.7 Conclusion

While NaaS offers numerous advantages, including scalability, cost savings, and agility, it is not without its challenges. Organizations must carefully weigh these limitations—such as vendor dependency, security risks, performance variability, and integration complexities—before deciding to adopt NaaS. By understanding these challenges and proactively mitigating the risks, enterprises can more effectively implement NaaS solutions that align with their business goals and future network requirements.

NaaS is a transformative model, but its success relies on thoughtful planning, proper vendor selection, and careful management of the technical and operational challenges associated with this evolving networking paradigm.

Chapter 11:

Conclusion and Future Outlook

As we conclude this exploration of Network-as-a-Service (NaaS), it is essential to reflect on the critical concepts and benefits, emerging opportunities, and strategic recommendations for enterprises considering NaaS. Additionally, we will look toward the future of enterprise networking and the transformative role NaaS will play in shaping this landscape.

11.1 Recap of Key NaaS Concepts and Benefits

Throughout this book, we have examined the fundamental principles of NaaS, which represents a shift from traditional, hardware-centric networking to flexible, cloud-based service models. The key concepts and benefits of NaaS include:

- **Scalability**: NaaS provides enterprises with the ability to scale their network resources dynamically, allowing for quick adjustments based on demand without the need for extensive capital investment in physical infrastructure.

- **Cost Efficiency**: By adopting a pay-as-you-go pricing model, organizations can optimize their networking expenses, paying only for the resources they consume while eliminating the need for overprovisioning.

- **Flexibility**: NaaS offers a variety of service models—such as Connectivity-as-a-Service, Virtual Network Functions, and Managed NaaS—that can be tailored to meet specific business needs, fostering greater agility in operations.

- **Enhanced Security**: Many NaaS solutions include built-in security features, such as secure cloud connections and integrated security management, which are essential in an increasingly threat-prone environment.

- **Simplified Management**: By offloading network management to NaaS providers, enterprises can focus on their core business functions, benefiting from the expertise of specialized providers.

11.2 Economic Implications of NaaS

The adoption of NaaS has profound economic implications for enterprises and the broader market. This section explores how NaaS can impact costs, drive productivity, and reshape business models.

11.2.1 Cost Management and TCO Reduction

One of the most significant economic benefits of NaaS is its potential for lowering total cost of ownership (TCO). Traditional networking often requires substantial upfront capital investments in hardware and infrastructure. In contrast, NaaS operates on a subscription-based or pay-as-you-go model, allowing enterprises to:

- **Reduce Capital Expenditures**: Organizations can eliminate the need for large upfront investments in networking hardware, reducing the burden on budgets and allowing for reallocating resources to strategic initiatives.

- **Optimize Operational Costs**: NaaS enables organizations to align their networking expenses with actual usage, avoiding costs associated with underutilized resources. This model leads to more predictable budgeting and financial forecasting.

- **Minimize Maintenance Costs**: By outsourcing network management to NaaS providers, enterprises can reduce ongoing maintenance and upgrade costs associated with traditional networking infrastructure.

11.2.2 Enhanced Productivity

NaaS not only lowers costs but also enhances productivity within organizations. By streamlining network management and improving connectivity, businesses can:

- **Accelerate Time to Market**: The agility provided by NaaS allows organizations to deploy new applications and services faster, reducing the time it takes to respond to market changes and customer demands.

- **Facilitate Innovation**: With reduced management overhead and costs, organizations can invest more resources in innovation, driving the development of new products and services that can generate additional revenue streams.

- **Improve Employee Collaboration**: Enhanced connectivity and access to cloud-based tools foster better collaboration among teams, improving overall productivity and operational efficiency.

11.2.3 Shifting Business Models

NaaS is facilitating the emergence of new business models across various industries. Organizations are leveraging the flexibility of NaaS to explore innovative approaches to service delivery and customer engagement:

- **Subscription-Based Services**: Businesses are increasingly adopting subscription models for their products and services, similar to NaaS. This shift can lead to more predictable revenue streams and improved customer loyalty.

- **Pay-As-You-Go Offerings**: NaaS encourages enterprises to offer pay-as-you-go pricing for their services, allowing customers to only pay for what they use. This model can attract new customers who prefer flexibility and lower upfront costs.

- **Partnership Ecosystems**: NaaS fosters collaboration between businesses and service providers, creating ecosystems where companies can integrate complementary services to enhance their offerings and reach new markets.

11.3 Emerging Opportunities in the NaaS Landscape

The NaaS landscape is rapidly evolving, presenting a wealth of emerging opportunities for businesses willing to embrace these changes:

- **Integration with Edge Computing**: As edge computing gains traction, NaaS providers will have the opportunity to offer services that enhance performance and reduce latency for applications requiring real-time data processing.

- **AI and Automation**: The integration of AI and machine learning into NaaS platforms will enable more sophisticated network management, predictive analytics, and automated incident response, resulting in more efficient operations.

- **5G Adoption**: The rollout of 5G networks presents significant opportunities for NaaS providers to deliver high-speed connectivity and support the growing demand for IoT applications and smart city initiatives.

- **Sustainability Initiatives**: With a growing emphasis on corporate social responsibility, NaaS providers can develop environmentally friendly solutions that reduce energy consumption and carbon footprints.

- **Industry-Specific Solutions**: NaaS can be customized to address the unique challenges of various industries, creating tailored solutions that enhance operational efficiency and competitiveness.

11.4 Recommendations for Enterprises Considering NaaS

As enterprises explore the potential of NaaS, several key recommendations can guide their decision-making processes:

- **Assess Business Needs**: Before adopting NaaS, organizations should conduct a thorough assessment of their networking needs, identifying pain points and goals that NaaS can help address.

- **Evaluate Providers**: Carefully evaluate potential NaaS providers based on their service offerings, reliability, security measures, and customer support. Look for partners that align with your organization's strategic goals.

- **Start Small**: Consider piloting NaaS solutions on a smaller scale before full-scale deployment. This allows organizations to evaluate performance, identify potential challenges, and refine their approach without significant risk.

- **Monitor and Optimize**: Continuously monitor network performance and usage metrics to ensure that the NaaS solution aligns with evolving business needs. Be prepared to adjust resources and configurations as required.

- **Prioritize Security**: As with any cloud-based service, prioritize security by implementing robust measures to protect data and network resources, including encryption, access controls, and compliance with regulatory standards.

11.5 The Future of Enterprise Networking

The future of enterprise networking is poised for transformative change, with NaaS at the forefront of this evolution. As organizations increasingly adopt cloud-native architectures and demand flexibility, the traditional network model will continue to shift towards more agile, dynamic solutions. Key trends shaping this future include:

- **Decentralized Networking**: The growth of edge computing and IoT will lead to more decentralized network architectures, where processing occurs closer to data sources, resulting in reduced latency and improved performance.

- **AI-Driven Networks**: Networks will become more intelligent, leveraging AI for automation, predictive maintenance, and real-time analytics to enhance operational efficiency and performance.

- **Hybrid Multi-Cloud Strategies**: As organizations adopt hybrid and multi-cloud strategies, NaaS will play a crucial role in seamlessly connecting diverse environments, enabling flexibility and optimizing resource allocation.

- **Enhanced Security Frameworks**: As cyber threats become increasingly sophisticated, the integration of security within NaaS offerings will be paramount. Future networks will incorporate advanced security protocols and real-time threat detection.

- **Sustainability and Social Responsibility**: As organizations prioritize sustainability, NaaS solutions that focus on reducing energy consumption and supporting green initiatives will become increasingly important.

11.6 Conclusion

NaaS is reshaping the enterprise networking landscape by offering flexible, scalable, and cost-effective solutions tailored to the unique needs of businesses across various industries. As organizations embrace the opportunities presented by NaaS, they can optimize their network infrastructure to drive innovation, enhance operational efficiency, and improve customer experiences.

By understanding the key concepts and benefits of NaaS, recognizing emerging opportunities, and following strategic recommendations, enterprises can position themselves for success in an increasingly digital and interconnected world. The future of enterprise networking is bright, and with NaaS leading the way, organizations will be well-equipped to navigate the complexities of modern business while achieving their strategic objectives. The economic implications of NaaS highlight its potential to drive growth, efficiency, and innovation, ultimately transforming the way businesses operate in the digital age.

Appendices

Appendix A: Glossary of Key Terms

1. **Network-as-a-Service (NaaS)**: A cloud service model that provides network services over the internet, allowing users to access, manage, and scale networking resources without the need for physical hardware.

2. **Total Cost of Ownership (TCO)**: A financial estimate intended to help buyers and owners assess the direct and indirect costs of a product or system over its entire lifecycle.

3. **Return on Investment (ROI)**: A performance measure used to evaluate the efficiency of an investment, calculated as the ratio of net profit to the cost of the investment.

4. **Virtual Network Functions (VNFs)**: Software-based networking functions that can be deployed in virtual environments, allowing for greater flexibility and scalability compared to traditional hardware solutions.

5. **Pay-As-You-Go (PAYG)**: A pricing model where users pay only for the services they consume, rather than a fixed fee, offering cost efficiency and flexibility.

6. **Edge Computing**: A distributed computing framework that brings computation and data storage closer to the location where it is needed, reducing latency and bandwidth use.

7. **Multi-Cloud**: An approach that uses multiple cloud computing services from different providers to meet specific business needs, enhancing flexibility and reducing vendor lock-in.

8. **Hybrid Cloud**: A computing environment that combines public and private cloud services, allowing data and applications to be shared between them.

9. **Internet of Things (IoT)**: The interconnection of everyday physical devices to the internet, enabling them to send and receive data.

10. **Software-Defined Networking (SDN)**: An approach to networking that uses software-based controllers or APIs to communicate with the underlying hardware, enabling more flexible and efficient network management.

Appendix B: NaaS Use Cases

Use Case 1: Healthcare Telemedicine

Scenario: A healthcare provider implements NaaS to support telemedicine services, enabling remote consultations and secure patient data sharing.

Benefits:

- Improved patient access to healthcare services.

- Enhanced collaboration among healthcare professionals.

- Cost-effective infrastructure for supporting telemedicine platforms.

Use Case 2: Retail Omnichannel Strategy

Scenario: A retail chain utilizes NaaS to integrate online and offline channels, providing a seamless shopping experience.

Benefits:

- Real-time inventory management across all channels.

- Enhanced customer engagement through personalized experiences.

- Reduced IT costs by leveraging cloud-based solutions.

Use Case 3: Manufacturing IoT Integration

Scenario: A manufacturing company adopts NaaS to connect IoT devices on the production floor, enabling real-time monitoring and analytics.

Benefits:

- Increased operational efficiency through real-time data access.

- Predictive maintenance capabilities to reduce downtime.

- Streamlined supply chain management with improved data sharing.

Appendix C: NaaS Implementation Checklist

1. **Assess Business Needs**:

   o   Identify current network challenges.

   o   Define specific goals for adopting NaaS.

2. **Evaluate Potential Providers**:

   o   Research different NaaS providers.

   o   Assess service offerings, security measures, and customer support.

3. **Pilot Testing**:

   o   Start with a small-scale implementation.

   o   Evaluate performance and gather feedback from users.

4. **Integration Planning**:

   o   Plan how NaaS will integrate with existing systems.

   o   Ensure compatibility with current applications and workflows.

5. **Monitor and Optimize**:

o   Set up performance metrics to evaluate NaaS effectiveness.

o   Regularly review and adjust network resources as needed.

6. **Security Considerations**:

o   Implement security protocols to protect data and network resources.

o   Ensure compliance with relevant regulations.

References

- **Cisco Systems**. (2020). *Network-as-a-Service (NaaS): A Transformative Cloud-Based Model for Enterprise Networking*. Cisco Whitepaper. Retrieved from *https://www.cisco.com*

- **Gartner**. (2021). *NaaS Market Guide: Trends, Growth and Opportunities*. Gartner Research Report. Available at *https://www.gartner.com*

- **IDC**. (2022). *Network-as-a-Service: The Evolution of Enterprise Networking in the Cloud Era*. IDC Whitepaper. Available at *https://www.idc.com*

- **Amazon Web Services (AWS)**. (2021). *Networking and Content Delivery Solutions in the Cloud*. AWS Documentation. Retrieved from *https://aws.amazon.com*

- **Megaport**. (2021). *The Benefits of NaaS for Global Enterprises: Case Studies and Use Cases*. Megaport Blog. Available at *https://www.megaport.com*

- **Juniper Networks**. (2020). *Leveraging NaaS for Scalable and Secure Enterprise Networks*. Juniper Networks Solutions Paper. Retrieved from *https://www.juniper.net*

- **Frost & Sullivan**. (2021). *NaaS Adoption: Key Drivers, Challenges, and Market Opportunities*. Market Research Report. Available at *https://www.frost.com*

- **Microsoft Azure**. (2022). *Enterprise Networking with Azure Virtual WAN and NaaS*. Azure Solutions Documentation. Retrieved from *https://azure.microsoft.com*

- **VMware**. (2021). *Multi-Cloud Networking with NaaS: Best Practices and Implementation Strategies*. VMware Blog. Available at https://blogs.vmware.com

- **Equinix**. (2022). *Enabling Hybrid and Multi-Cloud Connectivity Through NaaS*. Equinix Whitepaper. Retrieved from *https://www.equinix.com*

- **Accenture**. (2022). *The Role of Network-as-a-Service in Digital Transformation Strategies*. Accenture Research. Available at *https://www.accenture.com*

- **Open Networking Foundation**. (2020). *NaaS and SDN: A Pathway to Cloud-Native Networks*. Open Networking Foundation Blog. Retrieved from *https://www.opennetworking.org*

- **Ovum**. (2021). *The Economic Impact of NaaS: Total Cost of Ownership and ROI Analysis*. Ovum Report. Available at *https://www.ovo.com*

- **Google Cloud**. (2021). *Optimizing Network Performance with NaaS Solutions on Google Cloud*. Google Cloud Documentation. Retrieved from *https://cloud.google.com*

- **451 Research**. (2021). *NaaS: The Future of Enterprise Networking in a Multi-Cloud World*. 451 Research Paper. Retrieved from *https://www.451research.com*